The Drone Revolution

How Robotic Aviation Will Change the World

Steven M. Hogan

DEDICATION

This book is dedicated to everyone who helped along the way.

Let's go make the future happen.

CONTENTS

ACKNOWLEDGMENTS

This book would not be possible without the support of my law firm. It speaks volumes that my colleagues believe enough in the future to serve the legal needs of a brand new industry.

We have built something great together.
The horizon is ahead.
The future is bright.

DISCLAIMER!

No part of this book is legal advice. I say that because it is important! Please do not make legal decisions based on anything in this book. Seek the guidance of a legal professional who can talk through the issues with you. That's the only responsible thing to do.

All good? Ok: let's get down to business...

CHAPTER 1:
WELCOME TO THE REVOLUTION

Tomorrow Begins Today

We are at a turning point in the drone industry. The technology underlying robotic aviation is progressing more rapidly than anyone expected. The business potential for high-powered sensors on affordable platforms is nearly unlimited.

The sky is truly the limit.

Robotic aviation will change everything we know about what it means to fly. This is only the beginning. No one knows what the end will be.

This book is about where we have been and where we are going. I have worked in this field for three years. A very short time compared to some. A lifetime compared to others. This book is about what our future holds.

Robotic aviation is still new enough that individual people can have a huge impact on where the industry goes. This means that you can help decide what comes next.

The public is aware that "drones" are out there. They don't know much more than that. Your voice matters. What we do next will make all the difference in how the Drone Revolution plays out.

This is the tipping point. Let's work together to make tomorrow happen.

Everything Up to Now

Commercial aviation is in the midst of a revolution. A robotic revolution.

Small, autonomous helicopters and planes are changing the definition of "aircraft," "pilots," and "aviation" itself.

Entrepreneurial energy is flowing into the field with breakneck speed. We are in the "pioneer days" of the commercial drone industry. No one knows how it will end.

Perhaps there is no ending. The robots are here to stay. Commercial "aviation" will never be the same.

The "D-Word"

I use the word "drone." It doesn't scare me. It shouldn't scare you. It's just a word. Try it out and see. Say it out loud: "DRONE!"

See? Not scary. Just a word.

There is a minor controversy in the unmanned systems world about whether the "d-word" should be retired. This is a fun debate, no doubt. But that's all it is: a fun debate.

The public uses the word "drone" to mean everything from a Predator flying over a battlefield to a DJI Phantom flying over a dog park.

These things are not the same!

Nevertheless, the name has stuck. And we are stuck with it.

The better move is to ignore fights over terminology. That "battle" is a sideshow. Our energy is better spent with making the future.

How did YOU get here?

Every superhero has an origin story. I often find origin stories more interesting than the "super person" fighting it out with their arch-nemesis. The journey from Clark Kent to Superman means more than Superman's fights with Lex Luthor.

The pioneers of the commercial drone industry all have their own "Clark Kent" origin. I've heard all kinds of stories about how people found themselves here. Some started out as pilots, some are technologists, some have spent years flying model planes. Some, like me, are drawn to this industry by its novelty and game-changing potential.

No matter where we came from, we are now partners in an aviation revolution.

I'll start this book with my story. It's polite to introduce yourself at the start of a conversation. I look forward to hearing your story, too.

CHAPTER 2:
HOW I GOT HERE

Two Thousand Twelve

I'm a lawyer. A litigator all the way down. I wasn't always, though. I started out teaching high school social studies in St. Petersburg, Florida.

My first love as a teacher was making complicated ideas easy to understand. If you can say something simply, then you really get it. If you over-complicate things, you might not know what you're talking about. That part of teaching was a blast.

You might ask, "why law school, then?" The main thing is that I was bored. Boredom is a slow death. After four and a half years I had nothing left. Law school was my way out.

I graduated from law school in 2010. I went to work as a litigator, then got good at tax litigation. I still do those things and I like them very much. (Litigation is the *opposite* of boring!)

In January 2012, I had a conversation that would change everything. I was called in by one of the attorneys at my firm to talk about new, game-changing technologies. We thought that writing an article about an emerging technology would be a way to generate new business opportunities.

The theory was simple: find a new technology that's creating a new industry with unique legal needs. Then we go and meet those needs. I was on board immediately. That kind of open-ended challenge is right up my alley.

We made a short list of emerging technologies that looked like good options. "Domestic Use of Commercial Drones" was on our list.

FMRA and Fate

For the life of me I can't remember the other technologies we discussed. I wish I'd kept my notes! Circumstances quickly put us on the "commercial drone" path for good.

The President signed the FAA Modernization and Reform Act of 2012 in February of that year. That happened less than a month after our first discussion. We treated the FMRA as Fate and went all-in on the drone industry.

Law Clerks and White Papers

The first step was to puzzle through the current legal structure that commercial drones would operate in. I found that things were complicated, unclear, and changing quickly.

None of those things have changed. Today it's just complicated in a *different way*.

We are lucky enough to be located in Tallahassee, Florida. Not just because the city is a gem (it is!), but also because the Florida State University College of Law is located a few blocks from our law firm. This makes it easy to hire smart, motivated law students to clerk for us year-round.

Our "2012" clerks did a heroic job supporting our initial push to understand the drone industry. With their help, we drafted a "white paper" on the state of the law and made some predictions about how things might change. I am in debt to the tireless law clerks who spent their spring and summer helping us figure it out. They get the credit for jump-starting our work in this field.

That white paper gave us a way to introduce ourselves to the players in the drone industry. That was the beginning of the beginning.

Finding the Survivors

Terminator 3 is an extremely forgettable movie. Except for one scene that's stuck in my head for years.

After Skynet destroys the world, John Connor finds himself in a bunker with military communications equipment that (somehow!) still works. He turns on the radio and hears other human voices calling out to find each other. This was the start of the revolution. The survivors discovering they weren't alone.

This is what finding fellow-travelers in the commercial drone industry has felt like.

Fortunately, we didn't have to fire up a radio after the apocalypse to find our community. We used Google instead.

We quickly discovered a cluster of drone companies that had started up in Gainesville, Florida. These start-ups grew out of the robotics, biology, and agriculture programs at the University of Florida.

We reached out to the academic folks at UF and the companies in Gainesville, most of whom were happy to talk with us. We found more allies at Embry-Riddle in Daytona Beach. To a person, everyone was excited and willing to help.

These seeds of connection introduced us to a huge network of people in the drone industry in Florida and around the world. Over time these people have become our allies, co-travelers, and friends.

The network keeps growing. It's grown enough that you're reading this book.

There are always people willing to help. Google it. Find the survivors. Friends and allies are a phone call away.

Changing Hearts and Minds

It's critical for the drone industry to work together. The word "drone" is so politically fraught that lawmakers can easily pander to their constituents by "protecting" them from something new. Something scary. Scary because it's so new that it's hard to understand.

I also think the fear of drones is instinctual. Think about it for a minute. Surveillance satellites and photographs on manned aircraft can, and DO, take constant pictures of nearly the entire surface of the earth. Don't take my word for it – *go look at Google Maps.*

But those images have been accepted with a shrug by the public. I've never heard anyone say "OMG MY HOUSE IS ON GOOGLE MAPS PLEASE PROTECT MY PRIVACY!" Not once.

Instead, the response has been "OMG LOOK HOW COOL I CAN SEE MY HOUSE ON GOOGLE MAPS!"

The distinction doesn't make a bit of logical sense.

It makes emotional sense, though.

You can't see a satellite with your eye. You'd be hard pressed to see one if you had a good telescope. We're used to seeing manned aircraft flying around, and they are nearly always "way up high" – so far away that they don't seem real.

Telling a kid that a plane in the sky is an enormous jet is like telling them about Santa Claus. They'll believe you, but only like they believe in a night-flying "jolly old elf." Comprehension isn't the same as belief.

Contrast that with a consumer-grade drone. They fly low. You can see them. They sort of look like a bird. A hawk. A raptor. And now they are looking down on you. Taking pictures of you. You can't ignore them.

Why can't you ignore them? I don't think it's the novelty. I think that seeing an unfamiliar flying thing activates something deep in our lizard brain. Our fight-or-flight system kicks in and screams "THAT'S A PREDATOR! IT WILL EAT ME!" It makes sense. That kind of reaction is how we survived when our ancestors were as small as squirrels.

A squirrel is right to be freaked out by a hawk overhead. People can't help but react badly when they see something that triggers the "predator" response – and the name "PREDATOR" given to an infamous class of military-grade drones doesn't help from a P.R. perspective.

The antidote for all of this is education. Not the school kind, but the marketing kind.

When people hear a rational explanation of the economic value of drone technology, the light bulb goes on immediately. They get it. All it takes is one conversation with someone that understands the potential for drones to change every industry you can think of.

There will always be holdouts. There are some folks that are still nervous about automobiles. The Luddite demographic never really goes away. It just gets smaller.

So what's the answer? In a word: Advocacy.

It's not enough for the drone community to "find the survivors" and make friends with like-minded people. The next step is for the community to reach out to the world and tell its story to the public.

AUVSI (the Association for Unmanned Vehicle Systems, International) is the largest organization joining the drone community together. AUVSI chapters are the best vehicle right now for advocating reasonable policy choices at the federal and state level. (I am not paid by or affiliated with AUVSI other than being a member. This is from the heart!).

There are AUVSI chapters all around the country. My best advice is to find the one nearest to you and join it. You can pick a chapter to be "affiliated" with when you join AUVSI as an individual member. Or, as my friend Vince Donohue did with the "Heartland" chapter in Illinois, you can start a chapter if you need one close by.

There are two AUVSI chapters in Florida. The members of each one have been instrumental in knitting together a "coalition of the willing and informed" to press for legislative change in our state. This has turned out to be critical. The greatest obstacle to the commercial drone industry is NOT the FAA. The bigger danger is state-level regulation of drone technology.

I'll repeat that: *federal regulation* will not stop the progress of the drone industry nationwide.

State regulation is the biggest obstacle to success. It is also the greatest opportunity to thrive.

State Regulation of the Unfamiliar

States have rushed in to regulate what they don't understand. This has not had great results for the commercial drone industry around the country.

The National Conference of State Legislatures has compiled one of the most comprehensive databases to date of state "drone laws." The common theme is that states are preemptively regulating what they do not understand. This comes from an impulse to placate (and pander to) fearful constituents.

It's sad, but understandable. It is easy to "mess up" things you don't understand. The worst thing that the drone industry could do would be to disengage and write off legislative action as irrelevant. To the contrary, the states are where <u>all the action is</u> with regard to commercial operations.

There's a saying that "all politics is local." At the national level, this explains the motives behind many of the "fights" that look abstractly partisan on their surface. The core of these disputes is really about lawmakers playing to their "base" in their home districts. Every politician wants to get re-elected. There's no reason to think through hard issues when you can simply "throw rhetorical bombs" and make the folks at home happy.

There is another dynamic to the local focus of political action. It is absolutely true that lawmakers that are *closer to you* have more of an impact on your life. Your city council deciding whether or not to build a dog park is a lot more relevant to your life than whether the president bombs some distant country. Federal politicians will rarely "move the needle" on your day to day. For better or for worse, this is true.

So what does this mean for drones? It means that state legislation looms much larger for the growth of the commercial drone industry than the FAA does.

This may seem illogical, but hear me out.

All the action right now is centered on the FAA. We are all waiting for the FAA to write the rules allowing commercial operations without special, case-by-case permission. We all know this.

But play this out a little. The FAA <u>will</u> finalize its rules. Then everyone will know what constitutes a "safe" commercial operation. Then everyone can fly!

Great, right? But what comes next?

State legislatures will have *absolutely free reign* to decide what commercial operations, if any, are allowed within their borders.

I'll say it again a little differently: while the FAA will define what a safe flight is, the states will have a lot to say (maybe <u>everything</u> to say) about what you can do with a drone in the context of an otherwise safe flight.

Digest that a minute. This means that even if your proposed commercial operation is permissible under the FAA's safety standards, your state legislature has the power to <u>outlaw the operation entirely</u>.

All politics is local. Your state will impact your business much more than the FAA ever could.

CHAPTER 3:
<u>FEDERAL REGULATION OF DRONES</u>

<u>Getting Clear on Terms</u>

Though the "d-word" doesn't scare me, we have to get clear on our technical terminology. The word "drone" has no fixed meaning. It's a pop-culture term. It means whatever the speaker wants it to mean. "Drone" is less than helpful as a descriptive word.

The term "Unmanned Aircraft System" is more precise. This is a defined term under the FMRA that means "an unmanned aircraft and [its] associated elements (including communication links and the components that control the unmanned aircraft) that are required for the pilot in command to operate safely and efficiently in the national airspace system." *FMRA*, § 331(9). An Unmanned Aircraft <u>System</u> (or "UAS") is more than just the airborne device. A UAS includes all components that make the device work. This means the airframe, its mechanical components (engines, rotors, whatever), the controller used by the Pilot in Command (or "PIC"), the communications system that allows the airframe to be controlled by the PIC, and so on. Every piece that makes the device "work" is part of the unmanned <u>system</u>.

The UAS category is subdivided into two general sections. The first refers to the small devices that are easily accessible by consumers. These are "Small" Unmanned Aircraft Systems. This term is also defined by the FMRA as "an unmanned aircraft weighing less than 55 pounds." *FMRA*, § 331(6). This term is abbreviated as "sUAS."

To qualify as an sUAS, the airframe must weigh fifty-five pounds or less, fully loaded on takeoff. The weight includes all components and the payload. So if your camera is super-heavy, you could "tip the scales" and no longer have an sUAS on your hands.

There is no special term for a UAS that is <u>not</u> an sUAS due to weighing over 55 pounds. Anything over that weight is still a "UAS" for regulatory purposes.

When the FAA finalizes its rules for operation of sUAS in the National Airspace System (or "NAS") (get used to acronyms – they are part of the lingo!), they may include special rules for "micro" – or super-small – sUAS. That may entail a new term for micro UAS as that set of rules evolves. As of now, all UAS that are 55 pounds or less are properly referred to as sUAS.

sUAS: Where the Commercial Action Is

sUAS platforms offer the largest economic opportunity that has ever existed for robotic technology. This is a bold claim and I believe it completely.

Here's why: with sUAS platforms and small, light, inexpensive sensor components, collection of data for a whole host of industries is now within reach. Some examples:

- Precision Agriculture: showing farmers what plants are stressed, are in need of water, where pests are, etc. so that they can act accordingly in real-time to increase yields.
- Wildlife surveys.
- As-built inspections (in 3-D!) for buildings.
- Public Safety: finding lost children and adults.

- Emergency Management: quick, safe reconnaissance of a disaster area to identify survivors and allocate live-saving resources accordingly.
- Bridge inspections – up-close images of bridge components conducted quickly, cheaply, and reliably to check the maintenance needs of each span.
- Power line surveys: use of specialized cameras to identify "flares" where electricity is leaking out of components that need to be replaced or repaired.
- Surveys of mining operations to generate immediate topographical maps of what the terrain looks like after an excavation or blast.
- Shoreline surveys of sea turtle nesting sites to provide an accurate count of where they are without disturbing the wildlife.
- Fire department use of sUAS with thermal imaging cameras that will show where the "hot spots" are within a building before firefighters enter the structure. Identification of survivors in need of rescue can be accomplished in this manner as well.
- Awesome video of any event you can think of (weddings, surfing competitions, 5K and 10K races, marathons, etc.).
- Promotional video and still imagery for real estate sales.
- Monitoring of large tracts of land for environmental studies and research purposes.
- Identification of vulnerable populations of endangered wildlife without the use of human monitors in manned aircraft.
- *Let your imagination run wild.*

This technology is truly game changing in ways we cannot predict. "Big data" can now influence any industry you can think of. These kinds of operations can all be done with sUAS weighing far less than 55 pounds.

Waivers and Waiting

As I type these words, there is only one way to operate commercially according to the FAA. You must receive a waiver specific to your proposed commercial activity.

There are two general ways to get these waivers. The first is the "old and difficult" way. This is applying for Special Airworthiness Certificates in the Experimental or Restricted Category (SAC-EC and SAC-RC, respectively). This used to be the only way one could fly a UAS of any size for commercial purposes. This is how oil companies were approved to fly in the Arctic in 2013.[1]

The "easier" way to get flying is through the "Section 333 Exemption" process. This process is an example of the constant change in this industry.

The Section 333 process only "opened up" in September 25, 2014, when the FAA granted six exemptions for use of sUAS on closed motion picture sets.[2] That is barely seven months from where I sit today.[3]

The Section 333 process is only open for sUAS platforms. The legal authority for the FAA to grant these exemptions is found in Section 333 of the FMRA. This section reads as follows:

SEC. 333. SPECIAL RULES FOR CERTAIN UNMANNED AIRCRAFT SYSTEMS.

(a) IN GENERAL.—Notwithstanding any other requirement of this subtitle, and not later than 180 days after the date of enactment of this Act, the Secretary of Transportation shall determine if certain unmanned aircraft systems may operate safely in the national airspace system before completion of the plan and rulemaking

---

1 *See* "One Giant Leap for Unmanned Kind," FAA Press Release, July 26, 2013 (noting the "Restricted Category" approval for Arctic operation of two sUAS platforms).
2 All six of these exemption applications were filed by the law firm of Cooley, LLP, in Washington, D.C. The industry owes a debt of gratitude to their pioneering work!
3 April, 2015, to be precise.

required by section 332 of this Act or the guidance required by section 334 of this Act.

(b) ASSESSMENT OF UNMANNED AIRCRAFT SYSTEMS.— In making the determination under subsection (a), the Secretary shall determine, at a minimum—

(1) which types of unmanned aircraft systems, if any, as a result of their size, weight, speed, operational capability, proximity to airports and populated areas, and operation within visual line of sight do not create a hazard to users of the national airspace system or the public or pose a threat to national security; and

(2) whether a certificate of waiver, certificate of authorization, or airworthiness certification under section 44704 of title 49, United States Code, is required for the operation of unmanned aircraft systems identified under paragraph (1).

(c) REQUIREMENTS FOR SAFE OPERATION.— If the Secretary [of the Department of Transportation] determines under this section that certain unmanned aircraft systems may operate safely in the national airspace system, the Secretary shall establish requirements for the safe operation of such aircraft systems in the national airspace system.

FMRA, § 333.

That language has been in place since February, 2012. So why the wait? Who knows – your guess is as good as mine.

The best answer I have is that the FAA has a very complicated job in front of it. Creating a system that works everywhere in the country for a brand new technology is a tall order. It's a task made even harder because the technology is constantly morphing with breathtaking speed.

The upshot is that we now have a process for the FAA to determine whether "certain unmanned aircraft systems may operate safely in the national airspace system" in a given commercial context.

So what does this process look like? The application itself takes the form of a letter to the FAA. The letter is a "legal and technical" document explaining how a commercial operator will operate safely despite non compliance with some (or all) of the Federal Aviation Regulations.[4]

The application should list the technical specifications of the sUAS at issue and the procedures to follow in case something goes wrong. This could mean systems failure, loss of communication link, and anything else that could hurt the platform's ability to fly. The application should also state the general parameters of the proposed operations and the credentials and training of the Pilot in Command.

At least, that's the process today. It could change tomorrow.

For example, the process changed <u>while I was writing this book</u>.

The FAA initially required each recipient of an exemption to apply for separate "Certificates of Waiver or Authorization" (referred to as "COAs") before actually flying. This means that an applicant had to: (1) file for an exemption; (2) receive the exemption; and (3) <u>then</u> apply for a COA for each commercial operation.

While not ideal, this was certainly a better process than trying to apply for SAC-EC or SAC-RC approvals for each commercial flight.

As this writing, there are over 700 exemption requests pending. The FAA has granted just over 100 of them. You see the problem: the FAA could anticipate each exemption holder filing <u>hundreds</u> of COA applications while the rules are under consideration. That means the FAA

---

4 These regulations are largely codified in chapter 14 of the Code of Federal Regulations.

would have to individually examine each one. This could have gridlocked the process and fatally over-stretched the agency's resources.

This may be why the FAA loosened things up on March 20, 2015. The FAA announced a new process for granting "blanket" COAs to applicants that receive exemptions. In the memorandum announcing this new policy. The FAA would grant a "blanket" COA to successful applicants that allowed commercial operations in more restricted parameters than applicants normally ask for. However, the blanket COA would allow plenty of room to operate commercially.

The terms of the blanket COA that the FAA announced are:

- Operations are confined to 200 feet AGL maximum;
- Operations must take place during daylight Visual Flight Rule conditions;
- Operations must be within Visual Line of Sight (VLOS) of the operator;
- The operator must issue a Notice to Airmen (NOTAM) at least 24 hours prior to the proposed operation; and
- The operations must remain at least:
  - o 5 Nautical Miles (NM) from an airport with an operational control tower;
  - o 3 NM from an airport having a published instrument flight procedure but not an operational control tower; and
  - o 2 NM from an airport, heliport, or seaport that has neither a published instrument flight procedure nor an operational control tower.[5]

If an exemption holder needs to operate *differently* than the blanket COA provides for, the operator can still file for a COA detailing the specifications for a particular flight. Given that the FAA will be processing fewer COA applications than they would otherwise be without the blanket

---

5 *See* "FAA Streamlines UAS COAs for Section 333," FAA Press Release March 24, 2015.

COA process, these applications should even out at a 60-day approval window or less.

The fluidity of this process revealed itself again on April 9, as the FAA announced that it would institute a "summary grant" procedure for Section 333 applications, and that private pilot licenses (as opposed to recreational or sport pilot licenses) are no longer required.[6]

The NPRM: A Valentine's Day Gift

The Section 333 process will not last forever. It will be the "way things are" for the next two to three years, though. That's plenty long enough to justify the investment in applying for an exemption.

The next "phase" of things will be different, though. The FAA released its long-awaited proposed sUAS rule on February 15. Right after Valentine's Day. The fifteenth was a Sunday. How in the world did that happen?

It all started with Steve Zeets logging on to regulations.gov to check on the status of his Section 333 Exemption application. An unexpected document popped up in response to his search. This document looked like a statement of the FAA's policy in regard to sUAS regulation. He quickly downloaded it to his computer and shared it with some friends, none of whom could find the document on the website. Apparently, the document was taken down as quickly as it was posted.

As Professor Gregory S. McNeal of the Pepperdine University School of Law writes, "what Zeets didn't know was that he may have been the only person in the world to have been in the right place at the right time, able to download the inadvertently uploaded document before it was taken down by an unknown government official."[7]

---

6 *See* "FAA Summary Grants Speed UAS Exemptions," FAA Press Release, April 9, 2015.
7 Gregory S. McNeal, *Leaked FAA Document Provides Glimpse Into Drone Regulations*, Forbes.com (Feb. 14, 2015).

Up until that day, the drone community was nearly unanimous in expecting very restrictive regulations from the FAA. Few, if any, members of the commercial drone industry expected the NPRM to come out any way but awful.

You can forgive the pessimism. There was (and still is) plenty of frustration in the community with how "slow" the regulators have moved. This is the result of conflicting incentives and lived realities. There is no "good guy" or "bad guy" in this story.

On one hand, you have a flood of entrepreneurial energy moving into the sUAS market. This has been fueled by the widespread availability of relatively low-cost platforms delivering high-performance results. Of course, you get what you pay for. A $15,000 system is much more capable of handling high-end data collection and analysis than a $500 consumer-grade model. The more expensive you go, the more resilient the system will be to software malfunctions and the like.

On the other hand, you have the FAA – an agency with a mandate to keep the national airspace system ("NAS") *safe at all costs*. The FMRA handed the FAA the job of finding a way to integrate autonomous, pilotless devices into the NAS in a way that guaranteed the safety of manned flight. This is a tall order – especially when the Congressional mandate did not come with a whole lot of "extra money" to allow the FAA to staff the matter up.

The FAA's reaction was to be methodical and deliberate in implementing each part of the FMRA mandate. This led to wild frustration in the sUAS community as the FAA missed the FMRA deadlines.

For example, the final sUAS rule was supposed to be released by July, 2013, according to section 332(b) of the FMRA.[8] That did not happen. The

---

8 Section 332(b) directed the FAA to publish the "final rule" within 18 months of the FMRA (18 months following February, 2012, is July, 2013). The NPRM was supposed to be released two months following the FMRA, with the final rule to be completed 16 months later. *See FMRA* § 332(b)(1)-(2).

NPRM did not come out until February, 2015, after Mr. Zeets found the document in question.

What happened next was thrilling, at least for me. I remember checking Twitter before bed on Friday, February 13 (wild life, right?). One of the "drone news" feeds that I subscribe to posted the document that Mr. Zeets found. I paged through the document on my phone. I could not tell whether it was legitimate or a fake. I went to bed without retweeting in case the document ended up being a fraud.

On Saturday morning, I took my daughters to gymnastics class. My kids are at the age where the oldest can go out on the "gymnastics floor" for her class. This means I can sit with the younger one after her class is over and read silly things on my phone.

When I fired up Twitter to see what was up, nearly everyone I followed in the drone industry was talking about Mr. Zeets' discovery. Bloomberg News was the first to verify its authenticity. Professor McNeal wrote series of rapid-fire articles on Forbes analyzing the document and confirming it was real.[9]

The noise built to a crescendo that Saturday. Each hour provided a growing certainty that the document was real, and the FAA was taking an extremely progressive view toward regulating sUAS operations. This was a watershed moment. No one expected the rule to look this good.

The FAA put out a statement later that day informing the public of an "announcement" to come via conference call at 10:00 a.m. on Sunday. That Sunday was February 15th.

So you know what the drone community did on Valentine's Day? Rampantly speculated online about what this all might mean.

On Sunday at 10:00 a.m., the call-in line quickly broke down with hundreds of people dialing through. There was no guidance on the FAA website about what to do or whether another line would open up.

---

9 My favorite of the series is the article referenced above in Note 7.

In the chaos, Twitter blew up again with the drone community sharing information in real-time about who was on and what might happen. It was then that Brendan Schulman, the grand master of U.S. drone lawyers, coined the term "dronerati" for the Twitter accounts focused on the drone industry.

Finally, one of the "dronerati" tweeted out the new call-in number. I quickly called in. This was between 10:00 and 10:30 that Sunday morning. I remember it clearly. I gave my kids some food and instructions to "go play" while Daddy was on the phone.

Not everyone who called in made it through. By some miracle I got on the call. There were over two hundred people on the line. The FAA announced that the NPRM was indeed coming out and that it would look as good as we all thought. The same day, the White House issued an Executive Order on drones outlining the federal policy on data collection by autonomous devices.[10]

This was a huge turning point in the discussion. The commercial drone industry finally had some hope that the regulators were going to get this *right*.

That was a great Valentine's gift. A great big smooch from the FAA.

Better than chocolate or roses.

The Biggest Opportunity of Our Generation

On that Fateful Sunday Morning call, a question was asked to the FAA spokesperson: "how long will it take for the NPRM to be finalized into a rule?" The answer came back: *two to three years*.

That's a long time. It's also the blink of an eye.

---

10 *Presidential Memorandum: Promoting Economic Competitiveness While Safeguarding Privacy, Civil Rights, and Civil Liberties in Domestic Use of Unmanned Aircraft Systems*, White House Press Release (Feb. 15, 2015).

If you've read this far, you know I've been in this game since 2012. That's three years of struggling alongside industry pioneers trying to make a go of it with commercial drones.

That seems like a long time, but it's not. The days may be long but the years are short. The NPRM will be finalized before we know it.

In the meantime, commercial operators have a clear path to legal operation through applying for Section 333 Exemptions. This process is getting faster and more clear every day. You absolutely can start a drone business, right now, and get flying. The door is open wider than it's ever been.

Companies that get their exemptions now have a head start on the rest of the industry. The exemption process means that the entire industry is a "gated community" you have to get permission to join. There is nothing that an individual company can do to change this.

Complaining about it doesn't help. It can make you feel better for a moment, but that's fool's gold. The better option is to think about how you can take advantage of the situation.

Apply now. Get your exemption. Start making your mistakes. Figure out your business model. Make money *now* while the FAA grinds toward a final rule.

The opportunity is here for the taking. Drone companies will *never* have this advantage again. Once the NPRM is finalized, the doors will be open to a huge competitive environment.

This will lead to more drone manufacturers entering the industry. The price for high-end platforms will drop. They may drop *really fast*. This means that if you get good at providing drone services (DaaS, or "Drones As A Service"), you will have your pick of hardware platforms to experiment with.

Think about the opportunity. There is a two to three year window where your competitors will largely be "frozen out" of the market. If you get an exemption, you can show customers that *you* have the "special ticket" to operate legally under the FAA's regulations, all while your competitors *don't.*

How many customers will want to purchase services from an "illegal" operator? Would you feel comfortable with a "pirate" drone operation flying over your daughter's wedding? How about an expensive piece of real estate?

This is a huge competitive advantage. Now is the time to act. Every day that you don't apply for an exemption is time that you are not building your business ahead of intense competition. This opportunity will never, ever, come around again.

What You Need For An Exemption

When you file an exemption application, what does that look like? What do you need? This is another issue full of constant change. Right now, here are some general guidelines:

1.   Are you a Pilot?
Or do you at least *know* a pilot? This is a necessary first step in gaining an exemption to fly your drone for profit. The FAA is requiring every exemption applicant to pledge that the Pilot in Command (the "PIC") of the drone in question will have *at least* a recreational or "sport" pilot's license. Thankfully, the FAA no longer requires a medical clearance.

Now, is a pilot's license *really necessary* in order to fly an sUAS? Of course not. I know that, you know that, and anyone who's watched their buddy with a pilot's license crash a quad-rotor knows that.

This is what the FAA requires, though. They have not budged on this point for <u>any</u> granted exemption as of this book's publication date.

Practically, if you do not already hold a pilot's license, it's probably not worth your time to get one just for your drone business. The window of opportunity for Section 333 Exemptions will only be open for three years, maximum. Spending the time and treasure it will take to get your license will add unnecessary overhead to your business.

Of course, if you want to get a pilot's license because flying is awesome, go right ahead! You can deduct that as a business expense on your taxes, so the cost will effectively be discounted. Talk to your accountant or tax professional![11]

The better way forward, in my view, is to state in your exemption application that you will ensure that the PIC for each commercial flight has at least a recreational or sport pilot's license. You can find these people, hire them to fly for you, and off you go.

You do not need to worry about hiring them as "employees" with benefits, etcetera. The cleanest way forward may be to strike a deal where the PIC gets a percentage of the money made from each flight. If all they are doing is flying the device and you are putting together the data product for the customer, then a 25% cut to the PIC is probably sufficient.

It all depends on what you negotiate, but I think you can do that deal. That would be $62.50 on a $250.00 photography job. Maybe round up to $100 for the PIC and charge $350 to the customer. Experiment with different options and see what works best.[12]

The main thing is to find a PIC you can trust that can handle your sUAS. Family may or may not be the best choice. This will be a good experiment in your ability to read people. If you do the work in finding customers and putting the data together on the back-end, your

---

11 I'm a tax litigator too, so of course I had to mention this!

12 Talk to an accountant or other tax professional about the paperwork necessary to formalize this arrangement. You may need to issue a Form 1099 for IRS purposes, depending on your arrangement.

business model is set.

Of course, if you already have a pilot's license then you're ready for takeoff!

2. Flight Parameters?

Though the space is still in flux and the FAA could change its mind, generally flights are allowed up to 400 feet above ground level (or "AGL") and within visual line of sight ("VLOS") of the Pilot in Command ("PIC").

The "blanket COA" referenced earlier comes into play here. The FAA has started granting blanket approval for flights that meet the parameters set out earlier in this book.

This is important, as now an exemption holder is not required to file a separate COA application for each flight. That means that your only "wait time" is the period between when you file for your exemption and when you get it. You won't have an additional wait time for the FAA to approve your COA *after* it has blessed your flight plan in the exemption process.

The terms of the blanket COA are more restrictive than the flight parameters the FAA is approving for most Section 333 exemptions. The practical result is that if you have a business reason to fly at 400 feet (as opposed to 200 feet, for example), you will need to file a separate COA application for that operation.

3. What About Public Entities?

Everything I've just said about the Section 333 process pertains *only* to private entities. Public entities, like State Universities, Police Departments and Sheriff Offices, and government-run Environmental Protection agencies, do not need a commercial approval in order to operate.

Instead, public entities can skip straight to the COA process. This is a shorter path to get up and flying.[13]

The requirements for the COA are the same, mechanically. However, the public entity will have to have some evidence that it is truly "public" under the federal aviation regulations, thus rendering the sUAS a "public aircraft" under the pertinent rules.

I like to use a letter from the general counsel of the entity in question certifying that the aircraft is a "public aircraft" under 49 U.S.C. § 40102. This is normally sufficient.

If for some reason your entity is oddly structured, you may need different kinds of proof. That is a "nitty-gritty" question you should raise with a professional assisting you in the process. *(This book is not legal advice, contact your legal professional, etcetera.).*

The World After the NPRM

The window of opportunity to take advantage of the Section 333 process will close when the sUAS rule is finalized. Once the sUAS rule is "on the books," then the regulatory landscape will look quite different. You can expect the following changes if the final rule looks like the NPRM[14]:

1.   No Special Approvals Needed
     Commercial operators will not need to seek special exemptions from the FAA before flying. This will lower the industry's barrier to

---

13 An exception to this rule is when a public entity is engaged in commercial activity. The most common example is when a public university is offering training to students in return for tuition payments. The "commercial" nature of this transaction is undoubtedly tenuous, but it pays to be safe rather than sorry. No reason to risk the FAA shutting down your operation.

14 You can read the entire NPRM on the FAA's website if you are inclined to. Your best bet to find the NPRM is to go to www.faa.gov and search for the "UAS" page. I say this because the FAA has regularly switched its "drone" URLs around to reorganize the information. As of the date of this writing, you can find the NPRM here: https://www.faa.gov/uas/nprm/.

entry for anyone who meets the operator and aircraft certification requirements.

2. Easier Operator Certification

The NPRM states that "operators" of sUAS systems will not need to hold any sort of pilot's license. Instead, the operators will need to pass an aeronautical knowledge test at an FAA-approved knowledge testing center. No one knows exactly what this test will look like at the moment. Suffice to say it will likely require knowledge of how airspace classes work and similar information.

There may be a "vetting" process for operators conducted by the Transportation Safety Administration. Again, what this will look like is unclear.

Operators that pass the knowledge test and vetting process will receive an "unmanned aircraft operator certificate" with an sUAS rating. This certificate will never expire, but operators will be required to pass a recurring aeronautical test every two years.

The minimum age for operators may be as low as 17 years. Expect younger competition!

The sUAS used in operations must be available for inspection by the FAA. Accidents must be reported to the FAA within 10 days of any incident that causes personal injury or property damage.

Operators will have to perform pre-flight inspections of the sUAS in question.

3. Aircraft Requirements

The sUAS that an operator uses will have to be registered with the FAA. This is a simple process at the moment. It costs $5.00 and requires a form to be sent in to the FAA. As of the date of this writing, you can find instructions on how to register your sUAS here (.pdf link).

The aircraft registration will result in an "N-Number" that you must display on your sUAS. This "aircraft marking" will have to be displayed in the largest "practicable manner." The exact size will differ based on the size of your platform.

An FAA "airworthiness" certificate will not be required. This is a plus! It means that you will not have to prove to the FAA that your particular sUAS platform is ready and able to fly. This will help operators that purchase "out of the box" systems ready to fly. You will not have to be a technical expert to get up and going.

The sUAS will have to be maintained in a "condition for safe operation." The meaning of this phrase will change based on the requirements of the system. As the rules get finalized, we may see more specific requirements for various system classes.

4. Operational Limitations

The NPRM as currently written requires operators to stay within "visual line of sight," or "VLOS." This requirement has sparked a great deal of controversy in the sUAS community. VLOS means what it says – operations must take place within "visual" line of sight without enhancement. This means within sight of the naked eye.

The technology is quickly outstripping the VLOS requirement. First-Person View ("FPV") flying is rapidly becoming one of the most engaging ways to fly an sUAS. FPV is a term that encompasses any technology that allows the pilot to see as the sUAS sees – a real-time video view of the flight. When this works, it's like the PIC is "on board" the sUAS and can react quickly to any obstacles in its way. The technology is not advanced enough, however, for the FAA to feel comfortable in FPV-only controlled operations.

This does not mean that FPV cannot be used. To the contrary, FPV technology is specifically addressed in the NPRM and will likely be "okay" to use as a supplement to naked-eye control under VLOS conditions.

There have been a number of comments submitted to the FAA that address the proposed VLOS requirement. These comments may change the VLOS criteria for sUAS platforms. We cannot be sure at this time how the VLOS rule, if any, will turn out.

Closely tied to the VLOS requirement is the FAA's stance on daylight-only operations. The NPRM as written will not allow any "night flights." Daylight, in this context, means sunrise to sunset as measured by the local time applicable to the operation.

The VLOS requirement also folds into the FAA's requirement of a 3-mile visibility minimum as calculated from the control station. Practically, this means that you can't operate in rainy or foggy conditions (not that you'd get very good data if you did, anyway).

The maximum altitude for flights will be 500 feet. This is higher than the "blanket COA" currently issued to Section 333 exemption holders, and the 400 foot operational ceiling that most applicants are seeking.

Airspace limitations will apply as well. These issues can get complicated. Your best bet for a detailed explanation of airspace classes is to review the FAA's handbook on the topic (.pdf link).

For our purposes, it's sufficient to say that the NPRM states the following with regard to different airspace classes:

- Class A (18,000 feet and above): Operations prohibited.
- Class B, C, D, and E (regulated airspace): Operations allowed with permission of Air Traffic Control.
- Class G (unregulated airspace): Operations allowed without permission of Air Traffic Control.

Airspace classifications change based on where you are. Some airspace is "special use" and some is "restricted." You should call your local FAA Flight Standards District Office for information specific to your location.

Your sUAS may also be restricted from operating over persons "not directly involved in" the operation. This may pose a problem for event photography using sUAS. A way around this could be to define persons "involved in" the operation as those who consent to the photography operation. This issue will be an emerging one as the rules are finalized.

A potential solution to this problem is contained in the NPRM's "microUAS" option. This may allow greater freedom to operate "very" small sUAS in Class G airspace <u>and</u> over persons not "directly involved" in the operation. The NPRM proposes that this category include sUAS that are no heavier than 4.4 pounds (or two kilograms). This is similar to Canadian regulations.

Finally, the sUAS will need to be inspected pre-flight by the operator. The operator must not conduct an operation if the operator knows of a physical or mental condition that would interfere with safe operation of the sUAS.[15]

The Big Heavies

The NPRM we just reviewed <u>only</u> applies to the sUAS world! The "big boys" of robotic planes (think: the ones that can cross an ocean...) are an entirely different story. It will be a much longer time horizon before "robo pilots" are conducting flight operations in the NAS, whether autonomously or not.

These kinds of operations are certainly coming in the future. That's a subject for another day.

---

15 The NPRM is a 195-page document. This summary is necessarily brief and incomplete. I encourage you to skim through the full NPRM to get a sense of how things may work out. Please refer to Note 14, above, for instructions on accessing the full NPRM.

CHAPTER 4:
WHAT HAPPENS NEXT

The future is up to us. We get to decide what happens next.

The way we go about deciding will take three general forms:
- Facts on the Ground;
- Political Advocacy; and
- Litigation.

Let's look at each one.

Creating Facts On the Ground

The technology is moving too fast for most people to keep up with. The newest generation of sUAS that are available as I type this are vastly superior to what you could buy last year.

As the platforms, sensors, controllers, and user interfaces get better and more affordable, more people will be flying sUAS in their communities.

This is a fact of life for new technologies. Moore's Law spares no industry! Widespread adoption of sUAS technology is inevitable. Just like the supercomputers we carry in our pockets.

When technology becomes more widespread, people get more familiar with it. When kids are flying quad-rotors around the park, the word "drone" takes on a very different meaning.

Every time you fly an sUAS safely, talk to friends about the potential of the technology, and show the commercial "use cases" to people in various industries, you are playing a huge role in creating facts on the ground about sUAS operations.

Your actions will change how people think and feel about sUAS technology.

That will change how politicians react.

Political Advocacy

The biggest danger to the sUAS industry is bad regulations. As I said earlier in this book, the biggest roadblock will not be the FAA. Instead, the thing you should worry most about is what your state legislature does.

The most important thing you can do is to join up with like-minded people to press for sensible drone regulations. Right now, your local AUVSI chapter is the best way to do this. There may be better options for commercial operators in the future. Today, AUVSI is your best bet.

It's not just about joining together and talking amongst ourselves. That's an important part, for sure. We have to know and trust each other before we can do anything else. After establishing trust, the next step is to make your voice heard in your state's legislative process.

This may mean hiring a lobbyist. This is an expensive option but quite worth it if your industry faces a mortal threat.

This might also mean being your own lobbyist. This is a necessary step even if you hire someone. You must develop relationships with your legislators so they understand your industry.

This is not hard to do. Most state legislators are pro-business regardless of political party. Don't believe partisan hype: <u>everyone</u> wants their state to prosper. The only way to do that is to help entrepreneurs succeed.

The case is easy to make. Use the <u>AUVSI Economic Report</u> for your state to put hard numbers behind your enthusiasm for the industry.

This will work.

Here's how to do it:
1. Contact a group of like-minded people in your industry (these can be members of your local AUVSI chapter, or people that you are friends with in your state).
2. Find out which members of the group are willing to sign a "group letter" letter to their state legislators.
3. Find out who your "local" legislators are. In most states, every local district will have at least a state representative and state senator assigned to it. Most states make it very easy to figure out who your legislators are. Google "who is my state representative" and you're off to the races.
4. Write a letter that can be addressed to the local legislators of every person in the group. You can expand this to every member of the legislature if you like.
5. The letter should say who you are and what the sUAS industry can do for your state. Refer to the AUVSI economic report to show how the industry can create jobs in your state.
6. Send the letter! You should send both a physical copy via U.S. Mail to your legislator's home office and an electronic copy via email (or fax).
7. Follow up: give the letter about a week to be received. Then have the group <u>call</u> each legislator to follow up. You can divide up the list of legislators among your group so each person is not stuck calling 50 offices. If possible, the person calling should be located in the district where the legislator resides.

8. Get a meeting! Your group can and <u>should</u> request sit-down meetings with your legislators. The most efficient way to do this is to have your group travel to the state capitol when the legislators are there. In Florida, this is during "session." If you have a lobbyist, he or she can help coordinate these meetings. Often, though, you can schedule these meetings yourself by calling ahead. Alternatively, you can have group members set meetings at the home office for each legislator. This may be less expensive as it avoids travel costs.

These are just suggestions. I'm sure you can think of better ideas! The point is to get engaged so that your legislators <u>know you exist</u>. When a face gets put on the industry (YOUR face!), it becomes much harder for a legislator to vote for bad regulations.

Litigation

Litigation is about fighting things out in court. This is rarely a pleasant process for anyone other than the lawyers. You should avoid it if possible!

Remember, I'm telling you this as a lawyer who <u>really likes</u> to litigate. This is advice you should take!

Sometimes, though, you are left with no option other than to fight. You might find yourself in a position where the FAA makes a policy decision that hurts your business <u>so badly</u> that litigation is the only way to stay alive. You might find yourself hit by a state regulation that would cripple your business unless you push back in court.

When those are your options, litigation becomes inevitable and necessary.

There are two cases so far that show the potential value in litigation. I refer, of course, to the cases of Raphael Pirker (or "Trappy"), and the Equusearch organization.

1.   Trappy Fights the FAA

37

Raphael Pirker is the first person to contest the FAA's strict regulation of "commercial" sUAS. Mr. Pirker was flying commercially around the campus of the University of Virginia. He received a $10,000 fine from the FAA as a result of his efforts.

His case was taken up by Brendan Schulman, a top-notch New York lawyer with a long background in flying sUAS platforms. Mr. Schulman successfully argued to an Administrative Law Judge ("ALJ") that Mr. Pirker was not flying an "aircraft" as contemplated by the Federal Aviation Regulations. Since the regulations apply by their terms to manned aircraft, there was no way the FAA could fine Mr. Pirker for "failing" to follow them.

The ALJ's decision sent shockwaves through the UAS industry. A widely held belief was that the decision "legalized" all UAS operations. That was not really the case – the decision only applied to Mr. Pirker and it was immediately appealed. Nevertheless, the decision made the industry aware that the FAA could be successfully "fought" on this issue.

The FAA appealed the decision to the National Transportation and Safety Board ("NTSB").[16] The NTSB eventually overturned the ALJ's decision. Mr. Pirker then settled with the FAA for a token amount.

This case was the first one to challenge the FAA's ability to hold UAS operators liable under the regulations governing manned aircraft. What it shows us is that UAS operators do have a point that they are different than manned aircraft. This case will be remembered for a very long time.

You can read a synopsis of the case at the website of Mr. Schulman's law firm.

---

[16] My law firm filed a brief in this appeal supporting Mr. Pirker's position (.pdf link). The brief was filed on behalf of Angel Eyes UAV.

2.   Equusearch Wins by "Losing"

On the heels of Mr. Pirker's successful challenge to the FAA's fine, the Texas Equusearch organization found itself embroiled in a similar conflict. Equusearch is highly regarded search-and-rescue organization that works to find missing persons in the United States and overseas. A hallmark of the Equusearch mission is using new technologies to help find missing persons more rapidly.[17]

Equusearch is a highly successful organization. Since it was founded in 2000, Equusearch has coordinated over 1,400 volunteer searches in 42 states and eight foreign countries. It has found over 300 missing persons alive.

You can imagine how useful drone technology would be to an organization like Equusearch. An sUAS with a basic camera payload can cover more ground in less time than search parties on foot. This leads to faster results and saved lives. Equusearch was a pioneer in using sUAS technology to help find missing persons. They began their sUAS-based efforts in 2005. These operations resulted in finding the remains of eleven deceased missing persons. These operations also included aerial reconnaissance in support of volunteers searching on foot.

Equusearch entered the litigation arena after receiving emails from the FAA instructing it to "immediately" stop using sUAS technology. The FAA stated that operating outside of a specific COA was an "illegal operation," regardless of the flight parameters.

Brendan Schulman took up Equusearch's cause just as he had taken up Pirker's. Equusearch filed a lawsuit in federal court challenging the FAA's ability to stop sUAS operations in the absence of a drone-specific rule.

The lawsuit resulted in a dismissal of Equusearch's challenge to the FAA's emails. Normally a dismissal means you've "lost." However, things were a little different this time.

---

17 I encourage you to read more about Equusearch at its website, http://texasequusearch.org/about/.

The court explained that the case was being dismissed because the FAA emails <u>were not valid cease and desist orders</u>. The court ruled that since no consequences could flow from the emails, there was nothing for Equusearch to challenge. This is why I say that Equusearch "won by losing."

The FAA then issued Equusearch COAs to commence with its rescue operations. You can find more information about these events on <u>the Equusearch website</u>.

This case provides a template for how to challenge a "cease and desist" order from now until the sUAS rule is finalized. I would caution you not to get too excited about this result for your business model, however. Different facts could turn out differently. Section 333 Exemptions are the only "legally safe" way to fly commercially.

<u>Choose Your Own Adventure</u>

"<u>Choose Your Own Adventure</u>" books were very popular when I was a kid. The idea was that when you got to the end of a "chapter," you had a choice of what your character would do next ("go to page 42 if you take the cursed necklace… go to page 54 if you follow the mummy into the cave…).

The story would profoundly change based on each of your choices. (And you could always flip back and make a "different" choice if you didn't like what happened!).

I think about these books when I ponder the state of the drone industry. Things are small enough right now that our choices resonate far beyond our immediate reach. The actions of each company will shape the course of the industry in unpredictable ways. Our choices will profoundly change the future.

This is the ultimate "choose your own adventure" story. The thing is, we can't "go back" and make different choices to see what happens.

It's all happening <u>right now</u>. There is no "tomorrow" to wait for. All we have is today.

I encourage you to take action and make the industry great. Your ideas matter. Your vision can be realized. Your impact will resonate far beyond anything you can imagine.

The future is bright.

Let's choose our own adventure.

POSTLUDE

If you enjoyed this book, I invite you to join me at www.robotic-aviation.com. That is my personal blog about the drone industry. I will continue to grapple with these issues as the industry comes into its own.

I invite you to comment on any articles you find engaging. I'm interested in your perspective! The only way we get better at this is by giving each other honest feedback.

I look forward to the conversation.

Let's work together to make something great.

ABOUT THE AUTHOR

Steven M. Hogan lives in Tallahassee, Florida, with his family and an ancient Pomeranian. Steven frequently speaks and writes about "drone law" and state and local taxation. Contact information for Steven can be found at www.robotic-aviation.com.